THIRD SPACEX STARSHIP ROCKET LAUNCHED

The Inside Story Of SpaceX Starship Lost, The Milestone Reached And Previous Launch Tests

By

Regina Sharp

Table of Contents

Introduction

Uncover the inside story of the SpaceX Starship lost—a moment that shook the foundation of space exploration, yet propelled humanity's quest for the stars with renewed vigor. We unravel the intricacies of the milestone reached, shedding light on the remarkable advancements achieved amidst the cosmic abyss.

But before we journey into the depths of this groundbreaking mission, we must first understand the path that led us here. From the inception of SpaceX's ambitious vision to revolutionize space travel, to the meticulous planning and rigorous testing that preceded the launch, each step of the journey has been a testament to human ingenuity and perseverance.

Let's go into the heart of the SpaceX Starship program, exploring the trials and triumphs of previous launch tests that paved the way for this historic moment. From the exhilarating highs of successful missions to the sobering lessons learned from setbacks, every endeavor has shaped the narrative of our cosmic exploration.

Together, we will unravel the truth of the third SpaceX Starship rocket launch, delving deep into the inner workings of this audacious endeavor. As we navigate through the unknown depths of space, let us embark on this journey of discovery, fueled by the boundless curiosity and indomitable spirit of humanity. Welcome aboard the SpaceX Starship—where the future of space exploration awaits.

History of SpaceX

SpaceX is an American aerospace company established in 2002, pioneering the age of commercial space travel. Under the leadership of entrepreneur Elon Musk, SpaceX aims to transform the aerospace sector, making space travel affordable and accessible. Based in Hawthorne, California, SpaceX made significant strides in space exploration, achieving several milestones.

The company introduced the Falcon 1 rocket, a cost-effective two-stage liquid-fueled craft, revolutionizing satellite deployment into orbit. SpaceX's innovative approach included the development of the Merlin engine, offering a more economical alternative compared to traditional engines. Notably, SpaceX prioritized the development of reusable rockets, challenging the conventional single-use model prevalent in the industry.

Despite initial setbacks, including a fuel leak and fire during its maiden launch in March 2006, SpaceX secured lucrative contracts, notably from the U.S. government. Recognition came in the form of winning a NASA competition in August 2006, aiming to develop spacecraft for servicing the ISS post-space shuttle era. Subsequent Falcon 1 launches faced challenges but culminated in SpaceX becoming the first privately owned company to launch a liquid-fueled rocket into orbit in September 2008, followed by a significant NASA contract valued at over $1 billion.

In 2010, SpaceX unveiled the Falcon 9, featuring nine engines, and commenced construction of a launch site for the Falcon Heavy, envisioned to lower the cost of transporting payloads to orbit. Notably, in December 2010, SpaceX achieved another milestone by releasing the Dragon capsule into orbit and successfully recovering it—a feat repeated in May 2012, when Dragon became the first commercial spacecraft to dock with the ISS, delivering cargo. Furthermore, in August 2012, SpaceX secured a NASA contract to develop a successor to the space shuttle for crewed missions.

The Falcon 9 was engineered with a reusable first stage, marking a significant advancement in space technology. In 2015, a Falcon 9 first stage achieved a successful return to Earth near its launch site.

Subsequently, starting in 2016, SpaceX introduced drone ships for rocket stage landings, further enhancing reusability. In 2017, a previously returned rocket stage was successfully reused in a launch, demonstrating the viability of the concept. Concurrently, a Dragon capsule was reused in a mission to the ISS, showcasing SpaceX's commitment to cost-effective space travel.

The debut test flight of the Falcon Heavy occurred in 2018, featuring successful landings of two out of three first stages, while the third made a controlled water landing near the drone ship. Notably, the Falcon Heavy's maiden voyage carried a unique payload—an orbit-bound Tesla Roadster with a mannequin in a spacesuit seated at the wheel. The first operational flight of the Falcon

Heavy took place on April 11, 2019, further solidifying SpaceX's position in the aerospace industry.

In 2019, SpaceX initiated launches for its Starlink megaconstellation, revolutionizing satellite Internet service. Each Falcon 9 flight carried approximately 50 Starlink satellites, contributing to the constellation's growth. By 2023, Starlink boasted 3,660 active satellites, representing half of all operational satellites in orbit. Additionally, the U.S. Federal Communications Commission approved an additional 7,500 satellites, with SpaceX aiming for a total of 29,988 satellites orbiting Earth at altitudes ranging from 340 to 614 km (211 to 381 miles).

On May 30, 2020, SpaceX achieved another milestone with the first crewed flight of a Dragon capsule to the ISS, carrying astronauts Doug Hurley and Robert Behnken. Concurrently, SpaceX unveiled plans for the Super Heavy–Starship system, intended to succeed the Falcon 9 and Falcon Heavy. This system, initially known as the BFR (Big Falcon Rocket), would feature a Super Heavy first stage capable of lifting 100,000 kg (220,000 pounds) to low Earth orbit. The payload, the Starship, serves multiple purposes, including rapid intercity transportation on Earth and establishment of bases on the Moon and Mars. Future missions include a lunar voyage in 2023 with Japanese entrepreneur Maezawa Yusaku and artists, as well as collaborations with NASA's Artemis program to land astronauts on the Moon, with ultimate aspirations of settling humans on Mars.

What is Dragon?

Dragon, a spacecraft crafted by the American corporation SpaceX, imprinted its name in history as the renowned pioneering private spacecraft to transport or ferry astronauts into orbit. Its inaugural test flight occurred on December 8, 2010, followed by a second test flight on May 22, 2012, which delivered cargo to the International Space Station (ISS). The maiden crewed flight on May 30, 2020, carried astronauts Doug Hurley and Robert Behnken to the ISS.

Featuring a bell-shaped pressurized forward compartment and an unpressurized cylindrical rear compartment known as the trunk, Dragon is equipped with solar arrays for power. Capable of transporting up to

6,000 kg (13,000 pounds) of supplies to the ISS, Dragon secured a contract with NASA for 20 uncrewed flights to the ISS by 2020.

The original Dragon model was succeeded by Dragon 2, offering two variants: Crew Dragon and Cargo Dragon. Crew Dragon accommodates up to seven astronauts and completed an uncrewed demonstration docking with the ISS in March 2019. The inaugural Crew Dragon 2 flight, Demo-2, launched in May 2020, followed by the fully operational mission, Crew-1, which transported four astronauts to the ISS in November 2020. Regular Crew Dragon launches to the ISS are scheduled every six months, while Cargo Dragon focuses on supplying the ISS.

Dragon launches atop the Falcon 9 launch vehicle from Cape Canaveral, Florida, and concludes its missions with a splashdown at sea.

SpaceX Third Launch

A SpaceX Starship rocket embarked on its third test flight from the Starbase facility in Boca Chica, Texas, achieving several milestones on Thursday morning before experiencing a probable breakup.

During the nearly hour-long integrated flight test, the deep-space rocket system aimed to conclude with a splashdown in the Indian Ocean, positioning the colossal vehicle for subsequent, more intricate test flights and, ultimately, missions to transport NASA astronauts to the moon's surface.

However, following reentry, the team encountered simultaneous loss of communication with Starlink, SpaceX's internet service, and with TDRSS (Tracking and Data Relay Satellite System).

Dan Huot, SpaceX communications manager, acknowledged the loss of the ship during the live broadcast, stating, "The team has determined that the vessel has been lost, therefore there will be no splashdown today," while emphasizing the progress made during the flight.

SpaceX routinely characterizes setbacks during early test flights as part of the development process. These flights serve to gather essential data, enabling engineers to refine Starship for future missions.

Consisting of the upper Starship spacecraft and the rocket booster Super Heavy, the Starship vehicle lifted off from SpaceX's private Starbase facility in Boca Chica, Texas, at 8:25 a.m. CT (9:25 a.m. ET).

The Super Heavy booster, the launch vehicle's bottommost part, ignited and ascended over the Gulf of Mexico. Although the booster separated from the Starship spacecraft as planned, it failed to ignite all expected engines for controlled ocean landing.

SpaceX is actively working to obtain video footage preceding the booster's water impact. Nevertheless, the booster surpassed previous flight milestones, as on prior flights, the Super Heavy booster was destroyed before attempting landing maneuvers.

The third test flight coincided with SpaceX's 22nd anniversary, as indicated during the live stream.

Aiming for Orbital Speeds

Elon Musk, CEO of SpaceX, has emphasized that the main objective of these initial test flights is to propel Starship to orbital velocities, enabling the spacecraft to achieve a stable orbit around Earth. Typically, achieving such speeds necessitates surpassing 17,500 miles per hour (28,000 kilometers per hour). While Starship successfully reached its goal of attaining orbital speeds during this flight, it did not intend to enter orbit.

Starship Tests and Tech Demos

Starship ignited its engine for approximately six minutes before transitioning into a coasting phase, during which the spacecraft underwent several crucial tests and technological demonstrations.

Initially, Starship accelerated to speeds approaching those required for orbital insertion. Additionally, the payload door—an essential component for deploying satellites into space post-orbit—was successfully opened and resealed, validating the mechanism's functionality.

The SpaceX Starship rocket system launched from Starbase in Boca Chica, Texas, for its third integrated test flight on Thursday.

Another significant test conducted by SpaceX was a "propellant transfer demonstration," aimed at transferring propellant between tanks onboard the Starship vehicle. This demonstration serves as an initial step in determining how Starship will be refueled during future orbital missions.

Following the achievement of several milestones, SpaceX made the decision not to reignite Starship's engines after the planned half-hour coasting phase during the flight test. This decision was made as Starship was on a steep trajectory, implying that Earth's gravitational pull would naturally cause Starship to descend, regardless of engine ignition.

Although the reason for SpaceX's decision to skip this test is unclear, engineers emphasized the need for thorough data evaluation in the coming hours and days.

During the livestream, it was highlighted that the Earth's atmosphere acts as a braking system for Starship, aiding in its descent.

Notably, the Starship spacecraft is equipped with approximately 18,000 lightweight ceramic hexagon tiles designed to shield it from the intense heat generated during re-entry into Earth's atmosphere.

As Starship re-entered the atmosphere, a vivid halo of bright red plasma—a result of extreme heat and pressure—surrounded the vehicle. Shortly thereafter, communication with the spacecraft was lost.

NASA Artemis Moon Mission

Ensuring the spacecraft is fully fueled will be essential for the success of Starship's upcoming significant missions.

During Starship's missions to the moon as part of NASA's Artemis program, it will necessitate remaining in orbit near Earth while SpaceX launches additional vehicles dedicated solely to fuel transportation. To reach the moon, SpaceX may need to conduct numerous refueling trips.

On Wednesday, SpaceX obtained regulatory approval to proceed with its latest test flight.

SpaceX's Explosive Test-Flight Process

Musk expressed greater confidence in the success of this flight compared to previous attempts in 2023. A successful outcome could yield valuable data, potentially enabling Starship to progress to more challenging test flights.

In a recent social media discussion, Musk cautiously estimated an 80% probability of reaching orbit, emphasizing the improved performance of the third flight's rocket compared to its predecessors.

However, SpaceX officials have consistently emphasized that they do not anticipate achieving 100% accuracy in these early test flights.

According to a statement on the company's website, each flight test serves as an opportunity for learning, conducted in a real flight environment rather than in a controlled laboratory setting. SpaceX's approach emphasizes rapid iterative development, which has underpinned the company's major innovative breakthroughs.

SpaceX Net Worth

Elon Musk's SpaceX is approaching a valuation of $180 billion as it prepares to offer insider shares at $97 each in a tender offer.

Ranked as the world's second-most valuable privately held startup, SpaceX has been in discussions regarding a tender offer ranging from $500 million to $750 million.

Originally deliberated at around $95 per share last week, the share price has since risen, reflecting strong investor interest in acquiring a stake in the leading space transportation company and its rapidly expanding Starlink internet-from-space service.

While details regarding the terms and final scale of the tender offer remain subject to potential adjustments, in August, the company secured an agreement with both new and existing investors to offer up to $750 million in insider stock at $81 per share in July, valuing the company at approximately $140 billion, as per a document from SpaceX Chief Financial Officer Bret Johnsen.

www.ingramcontent.com/pod-product-compliance
Lightning Source LLC
Chambersburg PA
CBHW070239260726
48658CB00006BA/2376